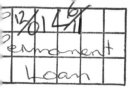

WAVES, TIDES AND CURRENTS

Daniel Rogers

The Sea

Exploiting the Sea
Exploring the Sea
Food from the Sea
Life in the Sea
The Ocean Floor
Waves, Tides and Currents

Cover picture: A magnificent wave on North Shore, Oahu Island, Hawaii.

Editor: Marcella Streets
Series Editor: Philippa Smith
Designer: Derek Lee

First published in 1990 by
Wayland (Publishers) Ltd
61 Western Road, Hove
East Sussex, BN3 1JD, England

British Library Cataloguing in Publication Data
Rogers, Daniel, *1955–*
Waves, tides and currents.
1. Ocean. Dynamics
I. Title II. Series
551.47

ISBN 1 85210 880 0

Phototypeset by Rachel Gibbs, Wayland
Printed and bound in Italy by L.E.G.O. S.p.A., Vicenza

CONTENTS

The calm waters of the Pacific Ocean gently lap the shores of Kiribati.

WAVES

If you were in a spaceship looking down at the surface of the planet earth, most of what you could see would be coloured blue. That is because more than 70 per cent of the earth is covered by water. Nearly all of that water is contained in four great hollows in the earth's surface: the Pacific, Atlantic, Indian and Arctic Oceans.

The water in these oceans, and in the smaller seas that are part of them, is never still. Although there is obvious movement when waves ruffle the ocean surface, water deeper down is constantly on the move, even when the surface looks calm.

Most waves are caused by the wind. But some, called tsunamis, are set in motion by underwater earthquakes or volcanoes. Waves are really rows of hills and valleys that move across the surface of the water. The top of a wave is called its crest and the valley between two waves is a trough. The vertical distance between a trough and a crest is known as wave height. Crests and troughs move through the water one after another. The horizontal distance from one crest to the next is called the wavelength.

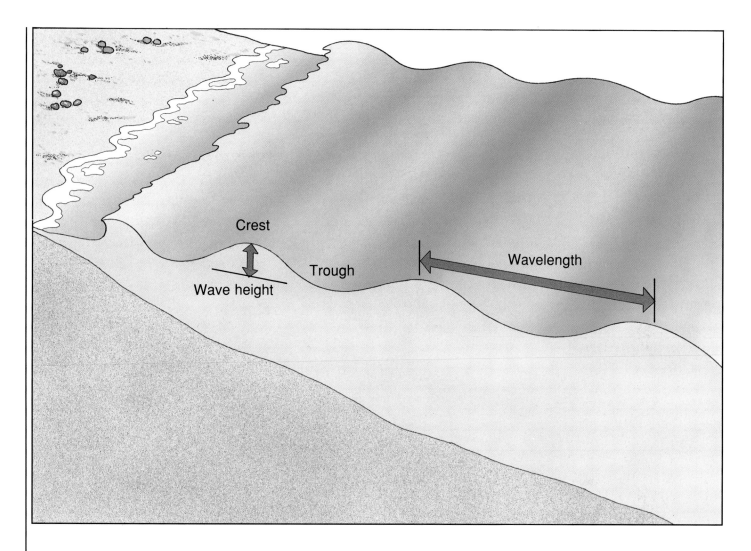

Crest

Wave height

Trough

Wavelength

When waves are moving in the open sea, they do not move boats and seabirds forward, but make them bob up and down. This is because the waves do not take the water along with them as they move; they actually pass through the water. Imagine a particle of water on the surface of the ocean. As a wave crest arrives it lifts the particle, then carries it forward, down and back. So the particle moves in a circle rather than being swept along. All of the water particles at the surface move round in circles as waves pass by, and this causes the particles lower down to move in smaller circles.

When waves reach a gently sloping shore they behave differently. In water less than half a wavelength deep, the wave crests become higher and steeper, and the wavelength gets shorter. The water particles then begin to move in oval shapes instead of circles, and those lower down catch on the sea-bed. This slows the wave down and makes the crest topple forward and break on the shore.

Above The height of a wave is measured from the trough to the crest. Wavelength is the distance between crests.

Left (top) Over 70 per cent of the earth is covered by water. *Left (bottom)* Even when the ocean surface is calm, the water deeper down is moving constantly.

WIND WAVES

Most of the oceans' waves are caused by wind blowing across the water. No one knows exactly how waves begin, but scientists think it happens like this: when the wind blows over the sea it seems to push down harder in some places than in others, and this makes patches of small ripples form; the ripples grow larger and larger, and when they are tall enough to be caught by the wind, they are blown along and become waves.

The wind moves waves in two ways. First, because wave crests stand up higher than the surrounding water, they are simply pushed along by the wind, rather like a sailing boat is blown across the water. Second, when the wind blows across the ocean it rubs against the water surface.

Waves like these on Waimanalo Beach, Oahu, Hawaii, are caused by wind.

This friction pulls the wave along. Because the wind has this double action, waves can sometimes travel faster than the wind that is driving them.

Wind waves can vary greatly in height, from a few centimetres to 12 – 15 m or even more. Nobody knows for certain how tall the biggest waves are in a powerful storm, but the largest wind wave on record was a massive 34 m high. It was seen in the Pacific Ocean during a strong gale. The height of wind waves depends not only on the speed of the wind but also on the length of time the wind is blowing, and on the distance over which it blows, known as the fetch. The largest waves are found in the Pacific.

This is because it is the largest of the world's oceans and so has the largest possible fetch.

When waves are formed in a part of the ocean where the winds are strong, they grow steadily in height, and their wavelength increases too. Eventually they may move out of the windy area where they were formed and into a calmer region. They then change into what is called a swell. This consists of low, smooth-crested waves. Swells can have a wavelength of up to 1 km, and they can travel over vast distances. Swells have been followed from the Indian Ocean, past Antarctica, through the Pacific and on to Alaska: a journey of over 19,000 km.

Because the Pacific Ocean is the largest in the world, it has the longest fetch.

THE POWER OF WAVES

The power of ocean waves is enormous. Every day several thousand waves pound against the coast, releasing as much energy as a large hydrogen bomb! This energy can cause enormous damage to coastlines, ships at sea and coastal towns and villages.

When fierce storms blow up at sea, the wind can create a complicated pattern of waves. The waves might be very different in size and may vary in wavelength, and travel in different directions. Some waves might grow up to 15 m high or more. They may get even bigger when they reach the shallow underwater shelf which surrounds the earth's continents. If a ship met one of these towering waves, it would probably either be destroyed by the enormous weight of water breaking over it, or it would plunge into the deep trough ahead of the wave.

When they reach the coast, waves can release their energy in dramatic and devastating ways. Breakwaters and lighthouses have to be designed to withstand the constant battering of the sea. In the early twentieth century, a storm sent huge waves crashing into the sea defences at Cherbourg in France. Part of the breakwater, a concrete block weighing 65 tonnes, was moved almost 20 m by the waves! One

Waves can be fun for surfers, but in fierce storms the power of the waves can destroy ships.

lighthouse on the coast of Oregon, USA, has to be protected from the sea with steel bars. Although it is over 42 m above sea-level, waves smash against the coast with such force that they can hurl rocks weighing 45 kg towards the lighthouse.

The immense power of the waves could possibly be put to good use. Scientists in Norway have built a device to channel waves into a small area so that their strength is concentrated. The waves' power lifts the sea water up into a high-level reservoir. When the water is allowed to run back down to the sea, it turns the blades of a turbine. The spinning turbine can then be used to generate electricity. Scientists in other countries are also trying to design wave power-stations. If they succeed, the sea could provide an endless supply of energy without causing the pollution problems associated with other fuels.

Lighthouses like this one at Portland Head, Maine, USA, have to be built to withstand the constant battering of the sea.

WAVES AND EROSION

The force of strong storm waves pounding against the coast can cause a great deal of damage. But even normal waves have the power to wear away the land. This is called erosion, and it happens in several ways.

When waves break against a cliff, it is not only water that hits the cliff face. The waves usually carry small rocks and pebbles which are thrown at the cliff as they crash against it. The pebbles wear away and break off pieces of the cliff. This type of erosion is called corrasion. Even grains of sand can have a corrasive action. The waves rub them against rocks like a giant piece of sandpaper. Corrasion happens mostly at the

base of cliffs, but boulders and small rocks can sometimes be thrown high up the cliff face during storms. The pebbles that wear away the cliff are also worn down as they are knocked against the cliff or each other. This process, called attrition, explains why most of the pebbles you see on a beach are smooth and rounded.

Waves also attack cliffs by hydraulic action. If you look at any cliff face you will see that it is not totally smooth. There will be cracks and holes in the rocks. Every time a wave breaks against the cliff, it traps air in these cracks

Above Caves like these in Dyfed, Wales, are created by waves. The sea attacks the cliff face, making small holes which eventually become caves.

Left This stunning cliff face on Loyalty Island in the Pacific Ocean was sculpted by waves.

Left *Old Harry Rocks in Dorset, England, have been eroded over many years. Each rock was once part of the land. Waves have eaten into the cliffs, hollowing arches like those in the large rock. Eventually the arches may meet. If the arch roof caves in, stacks of rock (like the one on the far right) will remain.*

and holes. The force of the water squashes this air, which pushes the rocks apart in order to find a way out. Then, when the wave falls back, the squashed air is suddenly released, putting more strain on the rock. Eventually the rock will be weakened and pieces will fall off.

Cliffs are also worn away by the dissolving action of the sea itself. Sea water is slightly acidic and it can slowly dissolve the rocks.

Some types of rock are harder than others. Softer rocks are usually eroded more quickly than hard ones. As erosion takes place, the sea eats into the base of a cliff, leaving overhanging rocks above. When the sea cuts in far enough, the overhang will collapse. This will give the waves even more loose rocks to throw at the cliff. If the cliff is very hard, the waves may attack small holes in it and hollow them out to form caves. If two caves form on opposite sides of a headland they may both be enlarged until they meet to form an arch. Then, if the roof of the arch falls in, a column of rock, called a stack, will be left standing in front of the cliff.

As the cliffs are worn further back, a rocky shelf is left in front of them. This wave-cut platform is what remains of the cliffs that have disappeared.

Left *This wave-cut platform in Blacksod Bay, on the west coast of Ireland, is all that remains of what were once tall, majestic cliffs.*

NEW LAND FROM THE WAVES

When a coastline is being eroded by the sea, loose rocks, pebbles and smaller debris are produced. Other small particles of rock and soil, called silt, are dropped into the sea by the rivers that flow into it. Most of this material does not stay below the cliffs or at the river mouth where it falls. Some of it is carried out to sea and some is transported along the coast.

Waves do not always approach the coast head-on. Often the wind blows them towards the shore at an angle. Yet when the waves have broken and the water flows back towards the sea it moves at right angles to the shore. Any sand and

This Spanish seashore is littered with debris deposited by waves.

Waves carry pebbles and sand along the coastline in a process called longshore drift.

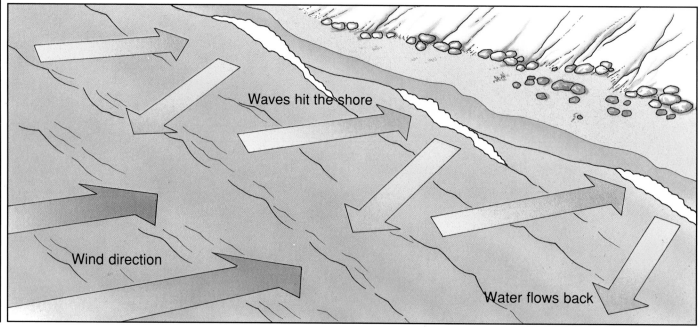

Waves hit the shore

Wind direction

Water flows back

pebbles being carried by the waves follow the same zigzag path. Gradually this process, called longshore drift, moves material along the coast.

The waves will continue to move material along the coast until an obstacle slows down the water. For example, a headland projecting out to sea will reduce longshore drift, and so will winds blowing against the direction of drift. When the water is slowed down, it cannot carry as much material as when it is moving fast. So some of the material is dropped, or deposited. If more material is being deposited than is being eroded and carried away, the deposited material is gradually built up to form new land. The most obvious examples of this are beaches, which may consist of anything from boulders to fine sand.

Deposition where the coastline changes direction, at a headland for example, sometimes results in ridges of sand and shingle, called spits, building up and growing out into the sea. Spits are often curved because the waves push them towards the shore. A tombolo is a type of spit that links an island with the mainland. Sometimes material is deposited a little way out to sea to form bars. These ridges form on gently sloping shores or across river mouths. They develop when waves break out at sea, dropping

the load they have been carrying. In time the bar may grow tall enough to be seen above the surface of the ocean, and may extend right across a bay from one headland to another. A bay which is cut off from the sea is called a lagoon. Bars are gradually pushed inland by the waves, and the lagoons become full of silt.

Top The bar of land on the left of this picture of Dalayan, Turkey, has been created from sand dropped by waves. The area of water right of the bar is called a lagoon.

Bottom Headlands reduce longshore drift.

TSUNAMIS

We have seen how powerful and destructive wind waves can be. However, the most dangerous waves of all are not produced by wind but by sudden disturbances of the ocean floor. They are called seismic waves or tsunamis, a Japanese word that means 'storm waves'.

There are a number of things that can bring about a dramatic shift in the ocean floor. An earthquake may cause one area of the sea-bed to drop down suddenly below the level of the area next to it, or be pushed up above its neighbour. A huge underwater landslide might send thousands of tonnes of rock crashing on to the ocean floor; a volcanic island or undersea volcano may suddenly erupt in a massive explosion. If one of these earth movements is strong enough, it will produce a series of waves spreading out through the water in all directions.

Unlike wind waves, which travel through only the surface waters of the oceans, tsunamis affect all the water from the sea-bed right up to the surface. This means that they are enormously powerful. In deep water they have wavelengths

Gigantic tsunamis destroyed this train at Seward, Alaska, following an earthquake in 1964.

of hundreds of kilometres and they can travel across vast oceans at speeds of 500 to 800 km per hour. Yet they are only 30 to 60 cm high, and can pass by a ship without anyone on board even noticing. But when they reach shallow water, a narrow bay or an inlet, their force becomes concentrated and they grow to a terrifying size. A wall of water, usually between 15 and 30 m high, sweeps on to the land causing immense devastation. Buildings and trees are flattened, crops destroyed, and people and animals killed.

Hundreds of tsunamis have been recorded during the last 2,500 years. One of the most famous happened more than a hundred years ago when the volcanic island of Krakatoa exploded. Krakatoa is in the Far East, between Java and Sumatra. In August 1883, three series of huge eruptions shook the area. They were so powerful that the sound was heard almost 5,000 km away. They were followed by massive tsunamis. When these giant waves swept ashore on Java and Sumatra they were up to 50 m high. Three hundred towns and villages were destroyed and 36,000 people drowned. Yet these were not the largest tsunamis ever seen. In 1964 an earthquake off the coast of Alaska, USA, produced one that was 67 m high.

A surfer rides a towering wave off the coast of Australia.

TIDES

Every day the level of the sea rises and falls. This rising and falling is caused by tides. At the coast the waves gradually reach higher up the beach. The highest point they reach is called high tide. Then they slowly slip back again to their lowest point, low tide.

Tides are actually like slow-moving waves that travel around the world's oceans. They are not very high but they have a huge wavelength. Unlike waves, tides are not caused by either the wind or by earthquakes, but by the moon. More than a thousand

The perfect scallop shapes created by waves on this Mexican shore indicate the high-tide mark.

years ago the Ancient Greeks noticed that tides were related to the monthly journey of the moon around the earth. But no one really knew exactly how they were related until Isaac Newton published his theory of gravitation in 1687.

As the moon orbits the earth, the force of gravity tries to pull them towards each other. Other forces stop them from coming together, but the moon's gravity is strong enough to attract the water in the earth's oceans. So, on the side of the earth facing the moon, the oceans are pulled outwards in a bulge. Meanwhile, on the opposite side, a second bulge is produced by the spinning of the earth, which throws water outwards. These two bulges travel around the earth's water surfaces, following the moon. The bulges create the high tides, and the low tides are the troughs that are between them.

If nothing else interfered with this pattern, every shore in the world would have two equal high tides and two equal low tides a day, as the earth makes one complete turn every twenty-four hours. But other forces do interfere. One of them is the sun. Like the moon, the sun also pulls on the oceans, but it has a weaker effect because it is much further away. However, when the sun, moon and earth are all in a line, the gravitational pull of the sun

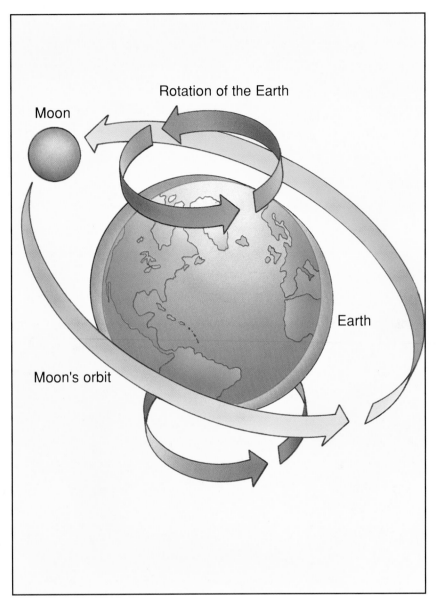

and the moon combine. This produces extra-high high tides and extra-low low tides, known as spring tides. When the sun, earth and moon form a right angle, high tides are very low and low tides are very high. This is because some of the moon's gravity is cancelled out by the sun. These are called neap tides. Each month there are two spring tides and two neap tides.

Tides are created by the moon. As gravity attracts the moon and earth towards each other, the water in the earth's oceans is pulled outward in a bulge. As the earth spins, the oceans bulge on the side of the earth opposite the moon. These bulges create high tides. The troughs between them are low tides.

TIDAL SYSTEMS AND TIDAL RANGE

It is not only the alignment of the sun, earth and moon that complicates the tidal pattern. Underwater ridges run across the floor of some ocean basins and these stop the tides from moving evenly across the whole ocean. Tides are also affected by the spin of the earth and by the drag of the water as it moves against the sea-bed.

The ocean basins are separated from one another by continents which form obstacles, preventing the tides from flowing right around the earth. This means that the pattern of the tides is not the same all around the world. In fact, even different parts of the same ocean can have different tidal systems.

Some shores have only one high and one low tide every day. They are called diurnal, or daily, tides. Others have semidiurnal tides: two high and two low tides each day. Yet others have a mixture of both: two tides per day but one very much stronger than the other. Most shores, including those around the Atlantic and Indian Oceans, have semidiurnal tides. The west coasts of Britain, western Europe and the coasts of the eastern USA follow this pattern, although the Gulf of Mexico has only one tide per day. Tides in many places around the shores of the Pacific Ocean are a mixture of diurnal and semidiurnal. But Alaska, far in the north, has diurnal tides. Some seas, such as the Baltic and the Mediterranean, which are almost completely cut off from the open ocean, have hardly any tides at all.

Just as the number of tides changes from place to place, so does the difference in height between high and low tide,

The shores of the Maldives in the Indian Ocean have semidiurnal tides: two high tides and two low tides each day.

known as the tidal range. In the open ocean this tidal range is quite small, often around a metre. But when high tide rushes into a narrow bay or inlet, it can increase dramatically. In the Bay of Fundy in Nova Scotia, Canada, the range for spring tides averages almost 14.5 m but it can be as much as 16.5 m.

There is a vast amount of energy locked up in the constant movement of the tides. Over twenty years ago French engineers built a new type of power-station that could harness some of this energy to produce electricity. They constructed a barrage, or dam, across the mouth of the River Rance. At the base of the barrage twenty-four large turbines are turned by the water surging past as the tides ebb and flow. The spinning turbines generate electricity. Similar tidal power-stations have been built in China and the USSR. Like waves, tides are a source of energy that will never run out, and they can be used to make electricity for us with a minimum of pollution.

The Black Sea has hardly any tides at all because it is almost entirely surrounded by land.

This barrage across the Rance estuary in France captures the power of the tides and uses it to generate electricity.

FLOODS

People who live by the sea are used to the rising and falling of the tides. Many have also experienced the minor floods that can happen when the spring tide is unusually high. But sometimes the tides can combine with severe weather conditions to make the coast a very dangerous place to live.

That is what happened in the North Sea at the end of January 1953. An area of very low pressure was centred over the sea. Winds swirled around it at 185 km per hour, whipping up huge waves. Unfortunately, this storm then coincided with a high tide. During the night of 31 January, with the tide more than 3 m higher than normal, the sea burst through coastal defences along the south-west coast of the Netherlands. Floodwater swept over a huge area of land, killing 1,835 people. Meanwhile, the east coast of Britain was also suffering severe floods. The Thames estuary and seaside towns in East Anglia were swamped and 300 people were drowned. The enormous waves also caused great damage to the coast itself. Along part of the Norfolk coast, a stretch of low cliffs was cut

back more than 27 m in just a few hours.

As we have seen, tsunamis can bring death and destruction on an even greater scale. Tropical storms – known in various parts of the world as cyclones, hurricanes, typhoons or willy-willies – also cause great damage and loss of life. One of the worst cyclones occurred in 1737 in the Indian Ocean. It hit the country we now call Bangladesh, drowning 300,000 people in one huge storm wave. It is often impossible to avert disasters of this kind. Occasionally, if there is enough warning, people can be moved out of an area which is threatened by a tsunami or tropical storm.

Coastal defences such as these tetrapods on the Danish coast can help prevent flooding. During storms they lock together tightly.

But frequently there are simply too many people to move in too little time.

Some types of coastal flooding can be prevented, however, by building strong sea walls and

breakwaters to protect the land behind them from storm waves and high tides. The port city of Osaka in Japan had been damaged by typhoon flooding so often that a complicated system of floodgates and pumping stations had to be built to protect the city. In London in 1984 a new type of sea defence was opened to prevent flooding from water surging up the River Thames during high tide and violent storms. The Thames Barrier, which extends across the river at Woolwich, has a number of large steel gates which can be closed when danger threatens.

Left The port city of Osaka in Japan has suffered from typhoon flooding so often that a system of floodgates and pumping stations had to be built to protect it.

Below The Thames Barrier in Woolwich, England was built to prevent London being flooded by storm waves from the North Sea.

CURRENTS

The oceans are never still. The winds blowing over them create waves and, on a larger scale, the influence of the moon causes the ebb and flow of the tides. There is something else that keeps the ocean waters constantly on the move: currents.

Currents are like giant rivers flowing through the sea. They are found both at the surface and deep down in the oceans, and they may be warm or cold, depending on where they have come from.

Surface currents are driven mainly by the great wind systems of the world, which blow with a fairly constant, steady force. In the warm tropics to the north and south of the equator, the trade winds push the oceans' surface water from east to west. Towards the far north and south, westerly winds drive the currents from west to east. But the currents have to change direction when they reach a continent. The spin of the earth also plays a part in deflecting the currents. North of the equator, currents turn to the right, and in the southern hemisphere they bend to the left. Deflection caused by the earth's spin is called the Coriolis effect, after Gaspard de Coriolis who discovered it. The result of all these factors is that the currents flow around the oceans in great loops called gyres. Scientists have found thirty-eight main currents in the world's oceans, and they form five huge gyres. For example, the North Atlantic Gyre is made up of four surface currents: the North Equatorial Current, the Gulf Stream, the North Atlantic Current and the Canaries Current.

Surface currents can contain a vast amount of water. A single current may carry more water than the biggest river on land. They usually move quite slowly – around 10 km a day – but narrower, faster currents are often found along the western edge of an ocean. The Gulf Stream is a warm current that

In the Caribbean Sea the warm currents of the Gulf Stream flow up to l60 km per day.

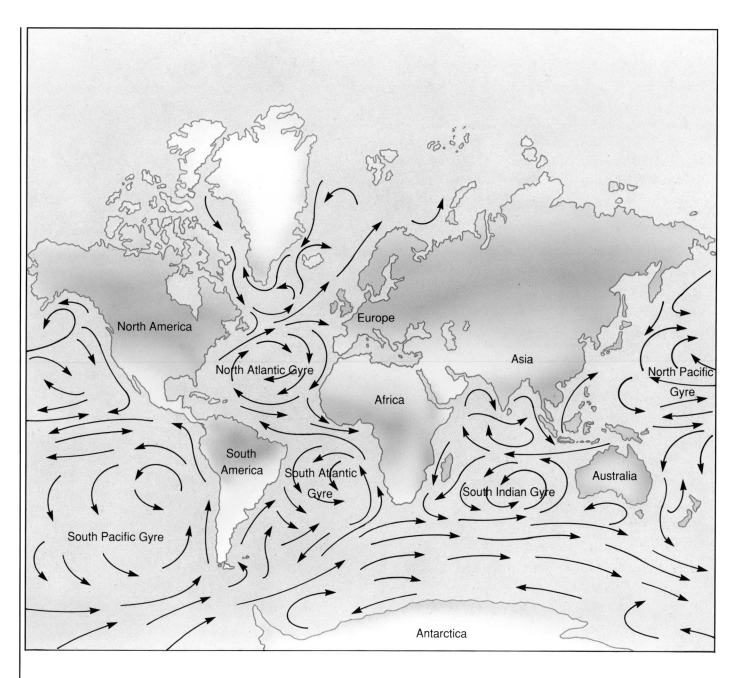

Labels on map: North America, Europe, Asia, North Pacific Gyre, Africa, North Atlantic Gyre, South America, South Atlantic Gyre, South Indian Gyre, Australia, South Pacific Gyre, Antarctica

flows from the Caribbean Sea and the Gulf of Mexico up through the West Atlantic. This warm river, 600 m deep and 60 km across, can move at up to 160 km per day.

As they move water around the oceans, the currents also transport heat. Warm, tropical surface currents like the Gulf Stream take heat away from the equator into colder regions to the north or south. They cool down as they go, and then turn back towards the tropics as cool currents. Meanwhile cold, heavy water around the poles sinks down to the ocean floor and begins its own long journey towards the equator.

This map shows the earth's largest oceans and the five main current gyres.

DEEP WATER CIRCULATION

Wind-driven currents only affect the water at the top of the oceans, from the surface down to about 800 m. Deeper down, in the darkness of the ocean floor, another system of currents is at work.

Deep ocean currents begin in the icy waters around the north polar ice and, in particular, around the continent of Antarctica. In the subzero temperatures found there, sea water freezes to form pack ice. But the salt in the water does not freeze with it, so the salinity – or saltiness – of the remaining unfrozen water increases. The higher the salinity, the heavier the water. This cold, heavy water

Deep ocean currents begin in icy waters such as those around Antarctica.

In the subzero temperatures of the Antarctic, sea water freezes into pack ice.

sinks down to collect on the continental shelf around Antarctica, and then spills over the edge and down the slope to the sea-bed. This dense mass of water is then swirled around Antarctica by the Antarctic Circumpolar Current and into the deep basins of the Atlantic, Indian and Pacific Oceans.

Currents are always strongest along the western side of an ocean. As they move into the centre and towards the east, their flow becomes less even as large eddies form. Deep ocean currents move much more slowly than surface currents. They may average less than 100 m a day, although scientists have recorded speeds of almost 5 km a day in some eddies. Special instruments are used to measure the speed of currents. Sometimes these current meters are attached to the sea-bed and left there for many months. When they need to

be recovered they can be released from the ocean floor by a remote signal from a survey ship on the surface above them. Alternatively they may be fixed to floats with weights to keep them at a certain depth in the ocean. The floats send out signals which can be picked up by ships and shore stations so that the current meters can be located easily. Another method is simply to hang a current meter on a very long wire from a survey ship and measure the current flowing past.

Special meters are used to measure the speed of currents.

CURRENTS AND CLIMATE

Ocean currents can have a very noticeable effect on the climate of places on or near the coast. One of the best examples of this is the Gulf Stream, in the group of currents that make up the North Atlantic Gyre. This current of warm water, at about 25°C, flows from the Caribbean Sea and the Gulf of Mexico up the east coast of the United States. It then splits and one branch, the North Atlantic Drift, flows north-east past Britain and Scandinavia. As a result, the climate of north-western Europe is warmer than other places which are just as far north. Some Norwegian ports which are well inside the Arctic Circle stay ice-free all year round because of the warm current. Other examples are the Kuroshio Current, which brings warm weather to the coasts of Japan, and the Brazil Current.

The influence of cold currents can be seen on the Atlantic coast of Canada. There the icy Labrador Current flows down from Greenland and the frozen

Ocean currents affect climate. Some Norwegian ports within the Arctic Circle remain ice-free all year because of warm currents.

Cold currents help create coastal deserts like this one in Western Australia. Cold water wells up from the depths of the ocean. When warm winds meet the cold water, fog forms at the coast but the air over the land remains there as a dry, stable mass that produces no rain.

north of Canada, past New-foundland and Nova Scotia. Although these coasts are actually further south than northern Norway, they have a cooler climate because of the Labrador Current.

Cold currents can help to create fog at sea. In the northern Pacific warm, moist winds blow northwards during the summer. Their path takes them first over the warm Kuroshio Current, and then across the cold Oyashio Current which flows south from the Bering Sea. The cooling of the warm, moist air over the Oyashio Current causes dense fog to form. This also happens off Newfoundland and along the coast of California.

Rainfall is also affected by cold currents. The Humboldt Current flows northward up the coasts of Chile and Peru in South America. It is caused by cold water welling up from the depths of the ocean. The warm winds that blow towards the coast are cooled down by the cold water, and fog forms. But the air warms up above the land and remains there as a dry, stable mass that produces no rain. This has helped to create one of the driest areas on earth, the Atacama Desert, and for the same reason deserts can be found along the western coasts of other continents. The Namib Desert on the coast of south-west Africa is there partly as a result of the cold Benguela Current. Other dry areas caused in this way are to be found in Lower California and on the west coast of Australia.

Even though it is further south than the ice-free coasts of Norway, Newfoundland in Canada has a colder climate. This is because of the cold Labrador Current.

LOADS THAT CURRENTS CARRY

We have seen how the ocean currents can move huge volumes of water from place to place and, at the same time, redistribute some of the world's heat. The moving water of the currents can also carry other loads.

In 1947 the Norwegian explorer Thor Heyerdahl demonstrated that currents may have helped to carry people from one part of the world to another. His balsawood raft, *Kon-Tiki*, drifted 6,900 km from South America to an island in the Pacific, showing that people in earlier civilizations might have used currents to make the same journey.

Currents also play a part in helping plants and animals to live. They carry food that ocean plants and animals need in order to survive. Some currents even carry the plants and animals themselves. Seeds that fall into the sea can be carried long distances by currents, and if they are washed up on another shore, they may take root and grow. The Gulf Stream transports baby eels, or elvers, that hatch out in the Sargasso Sea, a calm, weedy area of water in the Atlantic. The current carries some elvers to North America and others as far as Europe.

Cold currents flowing from the frozen north and from Antarctica in the south can carry icebergs

Below (left) Seeds that fall into the sea can be carried long distances by currents. If they are washed up on shore, they may take root and grow.

Below Cold currents flowing from the Arctic sometimes carry icebergs thousands of kilometres from the polar regions in which they formed.

thousands of kilometres away from where they first broke loose from the polar ice. As it flows southward, the Labrador Current brings from the Greenland icecap icebergs which sometimes drift into the North Atlantic Ocean. It was one such iceberg that caused the world's worst shipping disaster, the sinking of the great passenger liner *Titanic*. On the night of 14 April 1912, four days into its first voyage, the ship struck a huge iceberg in the Atlantic. At the time the *Titanic* was the largest ship ever built and it was thought to be unsinkable. Yet within two hours of hitting the iceberg, it lay on the ocean floor. Only 705 people were picked up by rescue ships; 1,513 drowned in the icy waters.

However, currents carry far more dangerous loads than icebergs. Some countries have dumped containers of dangerous nuclear and poisonous wastes on the sea-bed. This is no longer allowed, but if one of these containers were to leak, currents could then carry the deadly contents around an entire ocean and its shores.

This bird is a victim of pollution at sea. Its feathers are matted with sticky black oil.

GLOSSARY

Attrition The wearing away of pebbles when they are knocked by waves against each other or against cliffs, causing the pebbles to become round and smooth.

Beach A strip of land at the edge of the sea created by the waves depositing material they are carrying.

Continental shelf The underwater ledge that juts out into the sea at the edge of a continent.

Coriolis effect The force produced by the spin of the earth that deflects currents to the right in the northern hemisphere and to the left in the southern hemisphere.

Corrasion The wearing away of cliffs by pebbles and rocks carried by waves.

Crest The top of a wave.

Deposition The dropping of material being carried by the sea.

Diurnal tides Tides that have one high point and one low point each day.

Ebb To flow back.

Eddies Swirling movements caused when a current doubles back on itself to form a gentle type of whirlpool.

Erosion The gradual wearing away of the land by the sea (or by rivers, ice, wind and other natural forces).

Estuary The part of a river where it nears the sea.

Fetch The length of open water over which a wind blows.

Friction When one thing rubs against another; as the tides move across an ocean the water rubs against the sea-bed.

Gravity The force that attracts everything in the universe to everything else; the moon's gravity pulls on the water in the earth's oceans.

Gyre A loop of surface currents flowing around an ocean.

Headland A high piece of land which juts out into a sea or lake.

Hydraulic action The pressure of a liquid, such as sea water.

Iceberg A mass of ice which has broken away from the front of a glacier or the edge of an ice sheet, and is floating in the sea.

Longshore drift The movement of sand, pebbles and other material along the coast by the waves.

Low pressure Low pressure occurs when more air flows out of a region than flows into it. It often results in clouds and rain.

Neap tides Tides which have the smallest difference between high tide and low tide; they occur when the sun, earth and moon are at right angles to each other.

Pollution The spoiling of water, air or land by harmful substances.

Salinity Saltiness.

Seismic activity Earthquakes or earth tremors.

Semidiurnal tides Tides with two high periods and two low periods per day.

Solvent action Dissolving action.

Spring tides Tides which have the greatest difference between high and low tide. They occur when the sun, earth and moon are all in a line.

Swell Low, smooth-crested waves.

Tombolo A ridge of sand and shingle linking an island with the mainland.

Trade winds Strong, steady winds blowing from north-east to south-west in the northern hemisphere, and from south-east to north-west in the southern hemisphere. They are found in two 'bands' that run around the earth in the tropics.

Tropics   The area of the earth around the equator, between the Tropics of Cancer and Capricorn.

Trough   The lowest point in a wave.

Tsunami   A huge ocean wave set off by an underwater earthquake or volcanic eruption.

Westerlies   The winds that blow from the south-west in the northern hemisphere and from the north-west in the southern hemisphere. They are found in two 'bands' that run all around the earth, between the trade winds and the cold easterly winds that blow outward from the north and south pole.

BOOKS TO READ

Coasts by Keith Lye (Wayland, 1987)
Oceanography by Martyn Bramwell (Macdonald, 1988)
The Oceans by Martyn Bramwell (Franklin Watts, 1987)
The Oceans by David Lambert (Wayland, 1983)
Seas and Oceans by David Lambert (Wayland, 1987)
Water Energy by Graham Rickard (Wayland, 1990)

PICTURE ACKNOWLEDGEMENTS

Geoscience Features 20, 21 bottom, 25 top; Greenpeace *contents page*, 29 bottom; Robert Harding *cover*, 8, 15, 18; Institute of Oceanographic Studies 25 bottom; Japan National Tourist Organization 21 top; Frank Lane 13 bottom, 14, 16; National Trust 10 top, 11 top; Natural Science Photos 12 top (O. C. Rourke), 22 (Allan Smith), 28 right (Michael Chinery); Oxford Scientific Films 13 top, 19 bottom, 27 bottom, 28 left; PHOTRI 9 top and bottom, 27 top; Planet Earth Pictures 4 top (Flip Schulke/NASA) and bottom (Christian Petron), 7 (David E. Rowley), 10 bottom (Richard Chesher), 11 bottom (David George), 19 top (John and Gillian Lythgoe), 24 top (P. V. Tearle), 26 (Peter Stevenson), 29 top (John W. Warden). The artwork on pages 5, 12, 17 and 23 was provided by Peter Bull.

INDEX